STARTING OUT

baby ALLIGATORS

KIM THOMPSON

CREATIVE EDUCATION • CREATIVE PAPERBACKS

CONT

ENTS

I AM A HATCHLING.

I am a baby alligator.

I have about
75 sharp
teeth.

My mom laid about 40 eggs. She
covered them with grass and leaves.
I was ready to hatch. I made a peep.

My mom heard me! She
uncovered my egg.
I used a special egg tooth
to help me break out.

I have many brothers and sisters. We are a [pod](). Our mom protects us.

My mom
carries me in
her mouth.

Only my eyes and nostrils are above water.

I will get big! I may grow longer than a car and as heavy as a polar bear.

I sneak up on birds and other animals.

SPEAK AND LISTEN
UH-U

H·OW!

Can you speak like a hatchling? Baby alligators chirp and grunt.

Listen to these sounds:

https://www.youtube.com/watch?v=3WV69cAO1AY

Now it is your turn!

ALLIGATOR WORDS

hatch: to break out of an egg to be born

nostrils: openings in the nose that animals use to breathe

pod: a group of young alligators

scutes: hard, bony plates or scales that cover and protect an alligator or other reptile

READING CORNER

Scheffer, Janie. *American Alligators (The Ultimate Animal Library)*. Minnetonka, Minn.: Bellwether Media, 2024.

Smith, Emma Bland. *Claude: The True Story of a White Alligator*. Seattle, Wash.: Little Bigfoot, 2020.

Somaweera, Ruchira, and Stephanie Warren Drimmer. *The Ultimate Book of Reptiles*. Washington, D.C.: National Geographic Kids, 2023.

INDEX

PUBLISHED BY CREATIVE EDUCATION AND CREATIVE PAPERBACKS
P.O. Box 227, Mankato, Minnesota 56002
Creative Education and Creative Paperbacks
are imprints of The Creative Company
www.thecreativecompany.us

LIBRARY OF CONGRESS CATALOGING-IN-PUBLICATION DATA
Names: Thompson, Kim, 1970- author
Title: Baby alligators / Kim Thompson.
Description: Mankato, Minnesota : Creative Education and Creative Paperbacks, [2026] | Series: Starting out | Includes bibliographical references and index. | Audience term: juvenile | Audience: Ages 4-7 Creative Education and Creative Paperbacks | Audience: Grades K-1 Creative Education and Creative Paperbacks | Summary: "Introduce beginning readers to the world of baby alligators with this life science starter. Includes photos, a labeled animal diagram, "Make a Noise" section, glossary, and further resources"-- Provided by publisher.
Identifiers: LCCN 2024043235 (print) | LCCN 2024043236 (ebook) | ISBN 9798889897446 library binding | ISBN 9781682778302 paperback | ISBN 9798889897576 ebook
Subjects: LCSH: Alligators--Infancy--Juvenile literature
Classification: LCC QL666.C925 T484 2026 (print) | LCC QL666.C925 (ebook) | DDC 597.98/41392--dc23/eng/20250107
LC record available at https://lccn.loc.gov/2024043235
LC ebook record available at https://lccn.loc.gov/2024043236

DESIGN AND PRODUCTION
Design by Rhea Magaro
Production by Beeline Media and Design, Inc.
Art direction by Tom Morgan

PHOTOGRAPHS by Shutterstock/Elliott Cowand Jr, cover, Stile Art, 2-3, Robert Eastman, 4, Puffin's Pictures, 5, BEST-BACKGROUNDS, 6-7, Deborah Ferrin, 8, Svetlana Foote, 9, Heiko Kiera, 10-11, 14, Mark_Kostich, 11, Lindsey O, 12, Bohbeh, 13; Public Domain/FWC, 7

Printed in India